Running
Virtual
Meetings

20 MINUTE MANAGER SERIES

Get up to speed fast on essential business skills. Whether you're looking for a crash course or a brief refresher, you'll find just what you need in HBR's 20-Minute Manager series—foundational reading for ambitious professionals and aspiring executives. Each book is a concise, practical primer, so you'll have time to brush up on a variety of key management topics.

Advice you can quickly read and apply, from the most trusted source in business.

Titles include:

Creating Business Plans

Delegating Work

Difficult Conversations

Finance Basics

Getting Work Done

Giving Effective Feedback

Innovative Teams

Leading Virtual Teams

Other books in the series (continued):

20 MINUTE MANAGER SERIES

Running Virtual Meetings

Test your technology
Keep their attention
Connect across time zones

HARVARD BUSINESS REVIEW PRESS

Boston, Massachusetts

Copyright 2016 Harvard Business School Publishing Corporation

All rights reserved
Printed in the United States of America
10 9 8 7 6 5 4 3 2 1

No part of this publication may be reproduced, stored in or introduced into a retrieval system, or transmitted, in any form, or by any means (electronic, mechanical, photocopying, recording, or otherwise), without the prior permission of the publisher. Requests for permission should be directed to permissions@hbsp.harvard.edu, or mailed to Permissions, Harvard Business School Publishing, 60 Harvard Way, Boston, Massachusetts 02163.

The web addresses referenced in this book were live and correct at the time of the book's publication but may be subject to change.

Library of Congress Cataloging-in-Publication Data

Names: Harvard Business Review Press, issuing body.
Title: Running virtual meetings : test your technology, keep their attention, connect across time zones.
Other titles: 20 minute manager series.
Description: Boston, Massachusetts : Harvard Business Review Press, [2016] | Series: 20 minute manager series | Includes index.
Identifiers: LCCN 2016013072 (print) | LCCN 2016016130 (ebook) | ISBN 9781633691490 (pbk. : alk. paper) | ISBN 9781633691506 ()
Subjects: LCSH: Teleconferencing—Handbooks, manuals, etc. | Business Meetings—Management—Handbooks, manuals, etc. | Business Meetings—Technological innovations—Handbooks, manuals, etc.
Classification: LCC HF5734.7 .R86 2016 (print) | LCC HF5734.7 (ebook) | DDC 658.4/5602854678—dc23
LC record available at https://lccn.loc.gov/2016013072

ISBN: 9781633691490
eISBN: 9781633691506

Preview

Whether you're hosting a conference call, leading a WebEx meeting, or checking in with your team over an online chat, running an effective virtual meeting means overcoming an array of obstacles. Awkward silences, cacophonous interruptions, multitasking and disengaged participants, and all manner of technical glitches can cost you precious minutes. These are the problems you face daily as you work with people in different cities and countries. But you *can* prepare for and lead a productive meeting from afar. With guiding principles and tips for making your virtual meeting run smoothly, this book will have you ticking through every item on your agenda.

Running Virtual Meetings walks you through the basics of:

- Selecting the right channel for your virtual meeting

- Coordinating the technology that will allow everyone to connect

- Giving your colleagues the materials they need to meaningfully participate from afar

- Keeping participants engaged in the conversation, and bridging the in-person-versus-remote gap

- Assigning and clarifying meeting roles to keep things running smoothly

- Ending with a clear commitment to action for the whole group

- Holding participants accountable, whether they're in the office next door or on the other side of the globe

Contents

Contents

Running
Virtual
Meetings

What Is a Virtual Meeting?

What Is a Virtual Meeting?

L eading an effective virtual meeting is no easy task. There's the ghostly presence of other people on a conference call—a silent chorus that says hello at the beginning of the discussion and then never chimes in again with a question or an idea. The ding of an incoming e-mail from a teleconference participant who's not bothering to hide the fact that they're using the time to catch up on correspondence. Awkward, lengthy gaps as overly polite participants try to avoid speaking over one another. The colleague who takes a lagging video connection as license to interrupt other speakers whenever there's a pause in the transmission. And those are just the people problems. Slow internet service, software bugs, pixelated video—how do we make any progress on our agendas?

Problems like these can plague a meeting when you can't gather all the attendees in the same physical location. Whatever the purpose of your conversation, its dynamic will be fundamentally determined by the *medium* that brings everyone together. Video, conference call, messenger apps—as communication tools evolve, so do your options. Even if you work in a traditional office alongside your colleagues, you may spend most of your time in conference with people who aren't physically present: negotiating terms with the client who lives two time zones away; discussing a project with a colleague who regularly works from home; or just checking in with the coterie of collaborators who, like you, work out of coffee shops and libraries.

Though they're part of our regular workdays, these meetings aren't always as successful as they could be. There's a lot that can, and does, go wrong.

What are the key challenges?

You know what it feels like when an in-person meeting isn't going well. You're bored, you're frustrated, you're confused. Virtual meetings are susceptible to the same problems, but the causes are often different.

Technology. The success of a virtual meeting depends on the communication tools we employ. Picking the right technology is like picking a good meeting spot. You wouldn't ask colleagues to gather in an abandoned warehouse, at a members-only club, or at a restaurant an hour away from the office. Those aren't accessible locations, and your team won't feel safe or comfortable meeting there. Likewise, if your technology doesn't suit the group's needs, it can actually *prevent* people from participating. And even with the right tools, you'll have to deal with the usual run of problems, from dropped phone service to malfunctioning software.

Protocols. Meetings work best when people clearly understand the process and culture that they're expected to comply with. But when you're not face to face or in an office with established rhythms, your colleagues won't necessarily share your expectations and conventions. Who's moderating the conversation and has the right to recognize a speaker or ask someone to be quiet? What's the polite way to register disagreement—should they interrupt the flow of the meeting, or stay silent and follow up by e-mail later? As the leader of the meeting, you're responsible for clarifying these norms ahead of time and enforcing them as the meeting unfolds.

Engagement. Technology and protocols create a good space for virtual meetings; engagement is what happens in that space. But without in-person contact, your remote collaborators can easily become distracted or tune out. And if some people are attending the meeting in person and others are joining remotely, you face another set of difficulties. If you succeed in keeping the people in the room highly engaged, the virtual collaborators might be unsure how to participate in the group's vibrant dynamic and fall silent or feel excluded.

Why hold one?

If virtual meetings are so much trouble, why are they worth your time and attention? Because the way we're working requires them. As organizations extend their global reach, workers in diverse locations must work together to accomplish shared business goals. And as more and more cloud-enabled collaboration tools make their way to market, they offer a time- and cost-effective way to get that work done. If you need to assemble colleagues in London, Nairobi, and Mumbai, there are simply no other options.

But don't think of these meetings as your choice of last resort. Collaborate with people across town or an ocean away, and everyone reaps great benefits.

You'll build personal connections. For colleagues in different locations, virtual meetings are the only place to build rapport in real time. Whether you're using a video service or a chat app, these interactions let you engage in the small talk and rapid back-and-forth that's so essential to building trust—but so hard to accomplish over e-mail.

You'll take the onus for virtual collaboration off e-mail. We're accustomed to thinking about virtual meetings as serving the same functions as a "real" meeting, but they have another analogue: e-mail. For all its virtues, e-mail is not well suited for complex or technical discussions, and one-off questions or comments can get lost in the inexorable tide of new messages. Virtual meetings solve both these problems. They're a good forum for complicated conversations, and their low barrier to entry lets you initiate minor, but useful, interactions that otherwise wouldn't be worth the trouble.

You'll get more out of your time. You don't need to travel to your colleagues, or they to you. Carry out routine information-sharing without the hassle of a cross-country flight—or a ride up the elevator. Work with a corporate security specialist who has the skill set you need to navigate local regulations and infrastructure for the new office location you're planning. Invite your boss to address your team during its weekly hour-long meeting—then patch them in during the ten minutes they're available to talk.

What this book will do

Virtual meetings are challenging to facilitate, yes, but you can make progress on your work. Whether you're hopping online to meet with teammates in Japan or calling into the office on a work-from-home day, you can spark participation and engagement and lead better virtual meetings. *Running Virtual Meetings* is for you if you:

- Lead a virtual team

- Host meetings for the people you collaborate with, who work from various locations

- Give virtual presentations

- Have a direct report or colleague with whom you work closely and who works in a different location

To derive the most from these remote experiences, you'll need to prepare, conduct, and follow up on the discussion as you would for an in-person meeting. But as the leader of a virtual meeting, you need to:

- Plan your meeting, working around multiple time zones to schedule it and looping in the right people no matter where they are

- Choose the technology you'll use to host the meeting, and ensure it works

- Set—and enforce—expectations for participation

- Foster engagement, so that everyone contributes fully and feels ownership over the meeting's outcome

In the rest of this book, you'll learn practical tips for every step of taking those meeting basics and using them to run an effective virtual meeting. We'll begin by evaluating whether hosting a virtual meeting will help you achieve the goal you have in mind.

Plan Your Meeting

Plan Your Meeting

Running a successful meeting requires a fair amount of groundwork. Set yourself up for a meaningful, engaging conversation with the right players by thinking through your goal and deciding who can help you achieve it, what materials they'll need to contribute in an informed manner, and when and where you'll best assemble everyone.

Decide if you need to meet

Remote collaboration can exacerbate our tendency to plan needless meetings. Because we're measuring our working relationships against traditional office life, where coworkers see each other every day, we feel like we should be in contact with our counterparts more frequently than we are. Our team is dispersed, so group conference calls are our only way to interact, right?

But the cost of these interactions is invisible to us. We don't see how harried or bored our collaborators are during an ill-conceived meeting or how it can throw off the rest of their workday.

What are you trying to achieve?

To avoid wasting your colleagues' time—and your own—figure out what you hope to accomplish. Push yourself to be specific. For example, "My goal for this meeting is to make adjustments to our current project timeline. I want to gather input from the team about their schedule constraints, so that I can make a new timeline that people are able to commit to." As you're framing the goal, ask yourself:

- What do I want the outcome of this meeting to be?

- Why and how do other people need to be involved?

- Which people or points of view do I particularly want to hear from?

Are you ready to meet?

Now that you've articulated your goals, assess them critically. Will holding a meeting right now really accomplish what you want? Dig down on several key questions.

Do I have everything I need to discuss the situation? Do I have all the information necessary to make the meeting work? Am I clear in my own mind about my concerns and priorities—and those of the work? Perform a short thought experiment, where you put yourself in the place of key participants and imagine the questions they're likely to ask. Can you answer them? If not, postpone the meeting, and do some more research or independent strategic thinking.

Does moving forward require back-and-forth with collaborators in real time? If you simply need to get information from someone else ("What's your project status?" "Can you give me feedback on these documents?"), e-mail might be a better option, since colleagues can get to it on their own time. But if you need to discuss a complicated problem, get multiple perspectives on an issue, make a sensitive group decision, or do anything else that requires complex human interactions, plan for real-time talk.

Will our group dynamic benefit from a real-time interaction right now? Meetings matter for morale as well as efficiency. People feel more connected when they can see each other's faces or hear each other's voices. Even trading messages on chat can motivate, since the quick give-and-take builds rapport. If your collaborators seem disaffected or distrustful, try moving business you'd normally conduct over e-mail or through document sharing to a meeting.

After running through these questions, you may realize that you're not ready to bring in other people on this issue—or that you can take care of it on your own. But if this review confirms the need for a meeting, it's time to map out the plan to meet your goal, starting with a well-crafted agenda.

Create an agenda

Your agenda outlines the items you want to accomplish in a logical sequence, with a time limit and an owner for each order of business. As with a traditional meeting, break down complicated issues into component parts that build on one another. For example, if you need a plan for shifting team members' responsibilities, start by giving a relevant update (an agenda item might be "Review team members' roles on the project"); then get people's input on various ways to shift these roles ("Discuss potential approaches" and "Vote on approaches"); and, finally, make a decision ("Review votes, and decide on approach"). The lack of face-to-face stimulation shrinks people's attention spans, so plan the bulk of the meeting in 5- or 10-minute segments. If the meeting moves along at a good pace and people believe they're making progress, they're less likely to become distracted by their physical environments.

Factor in participant time zones and availability —if someone can join for only part of the meeting, schedule the items that matter to them in that window of time. Technology also influences your agenda. If attendees must use an onerous tool to view a presentation or access a database, group these activities together so that you don't lose momentum (or attention) as people log in or toggle between screens.

Include any information that folks will need to join the meeting on your agenda—where it's being held (conference room, chat channel), dial-in information, and so on. Add the backup plan details, too, so that participants have everything they need in one spot. Figure 1 shows a sample agenda.

A few days before the meeting, distribute your agenda with a packet of information about the meeting (see the section "Send Your Invitation and Materials" later in the book). To make sure people really read the information, pick a format they're likely to use. For example, if some participants are calling in from the road, make your packet mobile friendly.

Identify the participants

As you've been noting what topics you'll cover, you've probably also started jotting down people's names beside various items. To build the guest list, invite only

FIGURE 1

Sample meeting agenda

Agenda item	Who	Time allowed
Opening—quick overview of meeting roles and etiquette; review importance of Special Revenue-Generating Project and key team members' roles on the project	Emily	5 minutes
Review of team members' current responsibilities	Lisa	5 minutes
Discussion of possible responsibilities to shift away from team members	Lisa	10 minutes
Discussion of potential approaches	Lisa	15 minutes
Vote on potential approaches via Doodle poll	Group (Jack will share link for Doodle poll in team Slack channel)	5 minutes
Review of votes and final decision on an approach	Jack	10 minutes
Next steps	Emily	5 minutes

Purpose: Special Revenue-Generating Project: Reprioritizing Workloads

Objective: To develop plan for shifting other responsibilities of key team members

Attendees and roles: Emily, Lisa (facilitator), Jack, Chris (scribe), Angela (timekeeper), and Steve (tech czar)

Location: 5th floor conference room for NYC folks, WebEx for rest of team (information below)

Date and time: 2:00–2:55 p.m., January 25, 2016

Plan B: If we get disconnected or are unable to establish a quality connection, we'll move to the team Slack channel

Contact information for Steve, tech czar: Skype username swilde67

WebEx meeting information

Topic: Special Revenue-Generating Project: Reprioritizing Workloads

Date: Monday, January 25, 2016

Time: 2:00–2:55 p.m., Eastern Standard Time (New York, GMT-05:00)

Meeting number: 680 531 932

Meeting password: (This meeting does not require a password.)

To join the online meeting

Go to https://businessevents.webex.com/businessevents/j.php?MTID =mba123456789adbca3d156f170af9bc7

Audio conference information

To receive a call back, provide your phone number when you join the meeting, or call the number below and enter the access code.

Call-in toll-free number (US/Canada): 1-866-555-3239

Call-in toll number (US/Canada): 1-650-555-3300

Access code: 680 531 932

Global call-in numbers: If you're joining the conference from outside the United States or Canada, go to this website for dial-in information. https://businessevents.webex.com/businessevents/globalcallin.php? serviceType=MC&ED=123456789&tollFree=1

Toll-free dialing restrictions: http://www.webex.com/pdf/tollfree _restrictions.pdf

---

essential contributors. Attendees should fall into one of these four categories:

- Key decision makers

- People with information about what's under discussion

- People who have a stake in the issues

- Anyone who will be required to implement the decisions you make

The *number* of participants matters almost as much as their identity, especially in a virtual environment. On the one hand, communication technology allows you to gather large groups together with minimal logistical inconvenience. You don't need to book a massive conference room or pack your guests in cheek to jowl. You just need a platform that can support your audience size (some tools have a headcount cap). On the other hand, it's hard to moderate a discussion between static-filled voices or postage-stamp-sized faces on your computer screen. The following *8–18-1,800 rule* can help you finalize your invite list:

- If you have to *solve a problem or make a decision*, invite 8 people or fewer. If you have more participants, you may receive so much conflicting input that it's difficult to deal with the problem or make the decision at hand.

- If you want to *brainstorm*, your guest list can go as high as 18 people.

- If you need to *provide updates*, invite however many people need to receive the information. However, if everyone attending the meeting will be presenting, limit the number of participants to no more than 18 people.

- If you need to *rally the troops*, go for 1,800 people—or more!

Pick a platform

If your goals and your guest list fill the criteria for a successful meeting, it's time to decide how to hold it. And there's one piece of advice that experts agree on here: Use video.

Video

Sure, sometimes leading a virtual meeting using video technologies is impossible—when a coworker's calling in from their car, or when poor internet service on one end means that you'll all be subjected to a choppy connection. But where it *is* possible, you'll want to choose video over phone or chat for a couple reasons. Body language and facial expressions give you important information about people's reactions and moods. Everyone in the meeting will benefit from this extra layer of detail, and you'll be more likely to avoid misunderstandings and hurt feelings. And if you can see other participants' faces and peek into their lives a little bit, rapport and empathy will grow stronger: "Deborah's at an airport gate, and she looks pretty tired. This business trip must be grueling."

Video also discourages multitasking. It holds attendees accountable for their participation (yes, we see you making lunch), and the visual stimulation makes it easier to pay attention. When there's social pressure on all attendees not to appear distracted—and when you back up that norm by calling out bad behavior—your coworkers will be more engaged and your meetings more productive.

Nevertheless, video only works when all the participants have a strong, fast internet connection (at least 25 Mbps download speed and 3 Mbps upload speed). They also need a decent webcam—a stand-alone unit or one that's built into their device. And they need to be willing to turn it on. Some people feel deeply uncomfortable showing their work spaces to colleagues, and others freeze at the prospect of being on camera.

Other meeting channels

If you're working with people who can't engage with the content of the meeting because they're so unhappy about its format, you're better off choosing another medium. But which one? Broadly speaking, your options are phone (mobile, landline, or a tele-conference app like WebEx) or chat (via mobile phone or computer). Let's assess them.

Phone. Good for small-group discussions, simple information-sharing, and check-ins with small groups. It's your best bet for participants who lack reliable Wi-Fi or a place to set up their computers. Use it if you anticipate a lot of back-and-forth between partic-ipants, or if you're discussing something emotionally charged. If you share documents ahead of time, attendees can follow a presentation or review a report.

Chat. Good for any meeting with small groups, for casual check-ins with large groups, and for directed Q&As that require little discussion. It's convenient for people on the go, especially if you pick an application with a good mobile interface. But mobile keyboards will make it difficult for these people to participate in an extended conversation. This is not the ideal medium for emotionally charged material, although a less formal vibe may let people show their personalities a bit more, for example with emoji or GIFs. In general, user IDs help clarify the discussion, but without strong moderation, the sequencing of comments can quickly become unmanageable.

As a meeting platform, each of these tools has its drawbacks. You'll need to decide which user needs are the most important for your meeting and how to work around each mode's deficiencies. For example, if you're hosting a large meeting on a conference call, consider setting up a chat thread where participants can queue questions without interrupting the meeting. We'll cover setting up supplementary communication channels in the next chapter.

Set a time

Picking a time that works for everyone is especially tricky when you're in different time zones. Send around a poll (for example, with Doodle), or e-mail the team to find common availability. If your meeting is urgent, you might just need to wedge it in and hope people can rearrange their schedules to be there. But if you have a little wiggle room, find a time that works for the majority—and when people can really concentrate on the discussion. If you have access to invitees' calendars, check which days are packed and when major deadlines loom. Otherwise, ask up front about their schedules when you invite people: "I'd like to find a time to meet in the next two weeks to dedicate an hour to this issue. What looks best for you?"

You have a plan for when you'll meet and what you'll discuss with whom, but you're not ready to send your invite just yet. First you'll learn how to find the right combination of tools to connect and what to do if they don't perform the way you expected.

Manage the Technology

Manage the Technology

C hoosing video, phone, or chat is just the beginning. How can you determine whether your internet connection is actually up to hosting real-time file sharing between 14 people? Can you position your webcam so that people aren't distracted by the stacks of messy paper on the bookshelf behind you?

Even with something as basic as your meeting platform, much effort goes into making technology truly functional. Indeed, you probably rely on several other devices in addition to your main communication tool, from the tablet you use to view a PDF mid-chat to the Google Docs file you've created to capture real-time notes.

To make all this work, you'll identify which tools you need, practice ahead of time, and prepare for the inevitable technical glitches.

Assess your needs

"Your needs" is an umbrella term here. In fact, you should be thinking about three parties: you, your colleagues, and your organization (if you work for one). For yourself, start by considering how technology can support your own goals for the meeting:

- Will you need to record the meeting? If so, how? Will you take notes offline or on a document-sharing platform such as Google Drive, where participants can see each other's edits live?

- Will you require a secondary communication channel during the meeting? If so, will you use it to solicit questions or input from a large group, or do you want a private place to interact with individual participants (a presenter, your boss, a colleague) during the meeting?

- How will you share materials, such as an agenda or a presentation deck, before the meeting? Do you want people to be able to view and edit these documents together at the same time?

- How will you follow up? What do you want participants to do after this meeting, and which tools will they need to do that? How will you track their progress or post follow-up messages? If you expect the meeting to spill over into follow-up discussions, where will those take place?

Next, evaluate your attendees' needs. You're mainly concerned about accessibility here: Which setup will allow them to participate most fully in the conversation? Consider issues such as:

- Which meeting tools do they already know how to use? For tools that are unfamiliar, will they have time to learn how to use them before your meeting? Will you tutor new users or help them get the appropriate training on their own?

- Will they be using their computer, tablet, or phone?

- Will they be calling in from a car or staying put?

- Can they participate in a high-quality internet-based meeting? What results do they get from an internet speed test?

- How do they prefer to communicate? (For more on why this question matters, see the section "Bridge Linguistic or Cultural Barriers" in the last chapter.)

To answer these questions, gather input from the other participants, perhaps after drafting your agenda. If you hold your meeting on a platform that no one can operate, your meeting will fail. So if you decide to go with a new-to-them tool, talk to participants about how they plan to become familiar with it, and offer help if you can.

Finally, evaluate the security requirements your company has for this meeting. These rules protect your organization's business interests and shield it from legal liability. Planning your meeting in accordance with these rules helps your company manage risk and maintain its competitive advantage. Top security questions include:

- Do you need to record the meeting? What form should the recording take—for example, an auditory or a written transcript? Typed notes?

- Will you be discussing sensitive, proprietary information?

- What rules does your organization have for how that information is shared and stored? What security functionality does your meeting need (for example, with document sharing or a video connection)?

- Are participants in a location where they can safely speak about or view sensitive material? For example, do they have a quiet, private space (an office at another branch, a home office), or are they in a more chaotic and public location (a coffee shop, an airport lounge)?

Choose your tools

This exercise should generate a list of tool requirements ("I'll need to set up a chat thread for back-channel communications") and a list of constraints ("Katina can only join via mobile phone"). Map them against one another, and you'll come up with some viable options ("Text or a mobile-enabled chat service").

To pinpoint the best option, take an inventory of the resources currently available to you. Start by fixing your own location for this meeting. It could be your office, a conference room, or your kitchen table. What kind of setup can you reasonably manage from here? Pay attention to details such as video and teleconference links, a phone with speakerphone function, Wi-Fi signal strength, and the number of devices you can easily monitor at once. Then consider how each of these elements is likely to play out for remote attendees. For example, maybe broadband internet at your company's office means that you can run a teleconference app with complex security measures while streaming videos and using a cloud-based tool. But if your coworker is relying on hotspot technology at the airport, they probably can't keep up.

As you're working through these questions, find out what services IT support offers for setting up and running a virtual meeting. Can they suggest a tool based on your meeting's needs or talk your remote attendees through a software setup? Even if you don't have company employees on call to answer these questions, a onetime consult with a specialist might be worth the fee.

Test your setup

Once you've picked your tools, you'll want to try them out a few days before the meeting, to make sure you can start right in on your agenda on the actual meeting day.

Conduct a dry run. Tag one virtual attendee to test your meeting technologies. Schedule a time to practice initiating the conference call or video chat. Your checklist for this exercise includes:

- Making sure your colleague has all the information they need to join the meeting, including a call-in number, a link, or a passcode. If for some reason you can only share that information right before the call starts, make a plan for how you'll get it to everyone.

- Checking that the sound and video quality are good and that any lag time is manageable.

- Establishing that auxiliary tools, such as screen sharing or private chat, are working. Ascertain that your colleague can use these tools or view and receive materials during the call without losing audio or video quality or experiencing excessive delays.

- Putting in good order all the hardware you need, including cords and a power source.

If your setup is too elaborate—Skype plus Hip-Chat, with lots of screen sharing and open Google Docs files—you may want to scale back and then test your revised approach. Time how long it takes you to initiate the meeting, and decide whether you need an IT specialist on hand.

Confirm your space and resources before the meeting. The day before your meeting, double-check that the room you booked is really yours, that any hardware you're borrowing from the office is ready to go, and that the IT department knows what you need from it. At this time, you'll also want to send an e-mail with the meeting's call-in information to your attendees, and ask them to confirm that they have the space and resources to be productive participants. It's up to them to secure these conditions for themselves, but you can help with a timely reminder.

Do a preflight check. Get to the meeting location at least 15 minutes early to set up your materials and quickly test each function. If you're using a conference room, build this time into the room's booking.

What to do when technology fails

As the meeting leader, you have to anticipate and prepare for problems both on your end and with the other participants. Several best practices will help you handle any potential hiccups.

Appoint a tech czar. Ask one participant to act as the go-to problem solver for the other meeting participants: If someone can't access materials or if their internet connection keeps failing, they contact this tech czar. For this solution to work, take these steps:

- Pick someone familiar with the tools you're using, so the person can help solve basic problems. Don't pick someone who'll be presenting information or facilitating a discussion, since they won't have the bandwidth to do both jobs well.

- Provide a way for attendees to get in touch with them: phone, chat, e-mail, text—whatever works best for your meeting as a supplementary communication channel. But don't pick the medium that you're using for the meeting itself. Notify all attendees about this policy in the big e-mail you send out before the meeting.

- Invite the tech czar to sit in on your dry run for the meeting, and make sure they know how to get in touch with IT if they need additional help during the actual meeting.

Make a technology crisis card for yourself. Write down (on paper!) the tech information you might need if a problem arises during the meeting, and keep it next to your computer or tablet, so that you can quickly access it. Share a copy with your tech czar. Relevant items include:

- Name and phone number(s) of your IT support. Get the number for an individual specialist, not the departmental line.

- Name of your internet service provider and a help-line phone number.

- The account information for your most important tools, including the name of the account holder, the account number, the e-mail address it's registered under, password hints, purchase information, and so on.

- Name and version of your computer's current operating system.

You shouldn't actually be calling your internet provider's help line in the middle of a meeting—that's something IT support or your tech czar can do for you, while you carry on as best you can. But you want to hand off as much information as possible to these helpers so that they can work through the problem on their own, without interrupting the group to ask you questions.

Have a backup plan. What if you're unable to continue the meeting? In that event, be prepared to tell your colleagues what's happening, and decide how you want to proceed, whether that means rescheduling for later, moving to another communication channel, or canceling the meeting altogether and moving your business to e-mail. If you can make this backup plan ahead of time, do so. For example, you might send out a conference call number and tell people you'll switch to that channel in case of emergency.

Whatever approach you choose, don't leave your participants wondering what happened. Send an e-mail or a text explaining the situation: "The internet is down at our office, and we need to reschedule this meeting for a later date. Apologies for the inconvenience. I'll be in touch soon to find a time that works for everyone."

Canceling a meeting is frustrating, maybe even embarrassing, but it's not the worst thing that can happen with technology. What's worse is when your tools work, but everyone is using them in different ways. So how can you get everyone on the same page about how your meeting will work? We'll discuss that in the next chapter.

Set Expectations for Participation

Set Expectations
for Participation

Think back to the best meeting you've attended recently. What made it so successful? It probably shared some of these traits: There was an agenda, and the meeting leader stuck to it. Each person had a reason for being there: something to learn or to contribute. You all interacted appropriately, posing questions or sharing input without fighting for the group's attention or derailing the conversation. Everyone felt that the group processes were fair. For example, people saw decisions as legitimate, even if they didn't agree. The meeting accomplished something. You solved a problem, made a decision, generated ideas, or communicated information. And, the meeting finished on time.

Sounds great, right? But these things don't happen by chance. Group interactions work well when people behave well—that is, when everyone understands the rules and follows them.

When you're meeting without access to the information embedded in facial expressions and body language, you need to set clear expectations for everyone—including yourself—ahead of time. These standards needn't be absolute to be useful. There's more than one way to offer a dissenting opinion, for example, or to lead a discussion. What matters is that you communicate clear expectations—*before* the meeting. Call it your "code of conduct" or "meeting rules," and circulate it in the big e-mail you send out before the meeting to give everyone a chance to understand how they can participate productively.

Assign roles

No one can competently moderate a discussion, take detailed notes, mind the clock, *and* fix technical glitches. Fortunately, you have assistance: the rest of your attendees.

Assigning roles to meeting participants ahead of time allows you to communicate explicitly how you'd like each person to contribute to the meeting's goals. For example, if you want to draw out a quiet person, asking them to attend as an "expert" may give them the permission they feel they need to speak up about their specialization. To keep a chatty colleague from taking over, give them a task that will occupy their attention while making them feel important, such as timekeeper. In giving people a part to play, you give them a sense of how they should interact for the rest of the meeting. They won't be wondering, "What am I supposed to be doing on this call?" while they listen to disembodied voices argue about a technical point that's over their heads. They'll think, "I'm here to share my team's perspective when it's necessary, and mind the clock in the meantime."

You've already learned about one key role: tech czar. Participants can contribute in other ways.

Facilitator. Generally you'll assign this role to yourself. This person leads the agenda, solves problems, makes decisions, and keeps the discussion even-handed. The facilitator should be comfortable with virtual communication technologies and skilled at drawing out people who are less comfortable in these modes. Sometimes, it's beneficial to assign this role to someone else. For example, you may need to keep yourself on the sidelines during the discussion of a particularly sensitive agenda item, or you might want to give another attendee a leadership opportunity. If the meeting's attendees aren't part of the same organization or don't fall into an obvious hierarchy, appointing facilitators is a great way to share power, too.

Scribe. They take notes and share them, during or after the meeting. If you're using something more complicated than a pen and paper, however, they'll need to be familiar with your preferred note-taking technology. And if you're recording the meeting in your teleconference app, you may still want someone to jot down highlights—they're easier to share than the audio or video of an entire meeting. This role is a good way to engage quiet people.

Timekeeper. They track how long you spend on each agenda item and let you know if you're going over. Since they won't be able to catch your eye or make a subtle gesture from across the table, give your timekeeper the authority to speak up when necessary: "Sorry for interrupting—this is Ariel. We have fifteen minutes left in the meeting."

Presenter. Give these people a heads-up about your expectations well in advance of the meeting, especially if they need to do extra work to prepare. Later, check in with them about their presentation plans: What technology will they be using? Do they know how to present a slideshow within a teleconference app or how to talk the other attendees through a complicated spreadsheet over the phone? Offer to include them in your dry run if the presentation is complicated.

Expert. They bring a special skill set or useful experience to the conversation, one you want to make sure the rest of the group hears. If they don't need to be present for the rest of the meeting, consider asking them to call or log in at an appointed time. Either way, make sure they know in advance what input you'll want from them, so they can prepare accordingly.

Truth teller. They help you create an atmosphere of candor by modeling an honest engagement. Ask them to pipe up with contrarian points of view when a conversation becomes static and to call out inappropriate behavior: "It seems like a lot of us didn't read the premeeting materials. Next time, we need to do better." Or, "So, what's going on here that nobody's talking about?" Offer this role to an experienced colleague who knows the rest of the group fairly well, if that's possible.

Plant. They pose targeted questions to spark conversation. This role is especially useful if your meeting will gather many people who don't know each other well and who may be hesitant to speak up or ask questions. Sharing your slide deck or some sample questions with your plant ahead of time will give them the information they need to help generate a lively discussion.

Contributor. They're attending the meeting because they have generally useful knowledge and authority over the topic at hand, or they need information from the meeting to do their jobs.

Some roles—such as the facilitator or tech czar—require dedication and preparation, so set up a call to explain the role and to secure their enthusiastic agreement.

Establish meeting processes

The more participants and tools you're using, the more important it is to set expectations about how you'll conduct the meeting before you're actually in the meeting. Most traditional meeting practices apply, but some don't translate well to a virtual format. If you can't see someone's discomfort with a decision registered in their body language, how will you take the group's temperature before you move forward on the agenda? Here are the key dos and don'ts for helping your participants understand how you'll collaborate. Pick the ones that will work for your context and briefly add them to your meeting invitation and prep materials.

Do

- *Clarify how you'll make decisions.* Some issues you discuss may require a collective decision; others may fall to one or two people's judgment. Unless you're meeting with a team that already has a long-standing decision-making process, decide ahead of time and clarify up front about who gets to decide what, and who has veto power. You won't be able to look around the table and count votes. So plan how you'll solicit each person's opinion: a poll or survey? Orally, one by one? Prepare to keep a list of meeting attendees on your desktop to double-check. The person who doesn't speak up in the meeting might be a problem down the line.

- *Describe what kind of participation you want.* In face-to-face meetings, extroverts often have an advantage over introverts: They think out loud and feed off social interaction.

Introverts tend to hold back until they have a
fully formed point to make, and they may find
interactions with their colleagues exhausting.
Don't let technology worsen this imbalance. Let
the attendees know that you want to hear from
each of them, and that you'll call people out
individually when you want to hear more, or
ask them to step back from the discussion when
they start to dominate it.

Don't

- *Don't schedule status updates.* The social distance that technology introduces means your
  meetings need to be as engaging as possible.
  Don't force your colleagues to listen to each
  other recite project updates—they'll be bored,
  and you'll struggle to regain their attention. Instead, include this information in the premeeting reading material, or use another channel
  (such as a team wiki) to share them.

- *Don't let the group get off track.* Remote participants might miss body-language cues suggest-
  ing that their line of questioning isn't relevant
  to the group at large. And if other attendees
  see that you allow folks to get distracted by
  tangents, they might bring up their own unrelated issues and spark a discussion that leads
  you even further afield. Let everyone know that
  you'll be following the agenda carefully and
  that you'll address unrelated topics elsewhere.

Clarify etiquette

What will you do when two colleagues speak at once
or when one person's poor video connection arrests
the group's flow?

To some extent, the answers to these and related questions will depend on your particular situation: how well the participants know each other,
how formal or informal their work styles are, how

experienced they are with virtual collaboration. But considering possible scenarios and laying down some ground rules beforehand will help everyone get off on the right foot. Some suggestions:

- *Ban the mute button.* Tell your participants to have their microphones or telephones *on*. This tip is counterintuitive, but it comes straight from the experts. People go off task if they know that no one is listening—as far off task as a trip to the restroom. (In an InterCall study, 47% of respondents admitted to the latter vice.) In addition to holding attendees accountable for their presence, the inclusion of sound helps humanize everyone. Howling sirens on the street, the cooing baby—these noises break the ice and help people form a picture of who's on the other end of the line.

- *Prohibit multitasking.* You can't prevent colleagues from checking their e-mail or playing a game on their phone. But you can make it clear that such behavior isn't welcome and that you will call people out and follow up when someone's attention seems to be drifting.

- *Give a script for interruptions.* People are going to interrupt. Things like transmission delays and the lack of eye contact make phone, video, and chat fertile ground for this species of miscommunication. So tell people how they can politely interject. Do you want interrupters to hold their thoughts or ask for permission to go on? Will the facilitator manage who speaks by calling on people? Virtual technologies can make people's steamrolling tendencies worse, so develop a strategy for how you'll protect the speech of quieter folks. The section "Hear from Everyone" in the next chapter offers concrete suggestions for leading a balanced conversation.

- *Clarify how to use secondary communication channels.* You don't want attendees e-mailing or texting you with random observations ("Jill cut her hair!") or unproductive suggestions ("I'm not feeling this agenda"). So tell them what you *do* want to hear. How should participants solicit questions, comments, or updates? Provide tips for how participants should use other tools to solicit questions, offer comments, or give updates in the meeting. Or consider reserving secondary channels for emergencies only.

Send your invitation and materials

Now that you've decided on the participants, platform, meeting time, roles, and expectations, draft an e-mail invitation to your attendees with all the information they need to prepare for and participate in your meeting, including:

- Meeting time and time zone

- Your agenda, which addresses not only the items you'll discuss, the time allotted for each item, and who "owns" the item, but also the list of meeting roles (timekeeper, tech czar, and so on)

- Call-in or log-in information, including the tool's name, how to join, and the participant passcode

- Other technology you'll be using, including secondary communications channels, a document-sharing service, and authentication and security apps

- Contact information for the tech czar

- Details on plan B—which channel you'll move to if your primary meeting channel fails, and all the required information (call-in details, and so on) for this backup channel

- Guidelines for the appropriate usage of the tools you'll be using (that is, you're having a video call, but you'll use IM to let people know if they can't be heard, and so on)

- Ground rules for the meeting process and etiquette

- Any materials participants need to review before the meeting (presentations, sales reports, and the like)

Send this e-mail a few days ahead of time so group members have time to review the materials, vet their setup, and ask for help. For bigger meetings share the information even earlier, at least a week in advance.

The purpose of establishing rules and protocols isn't to intimidate your colleagues into compliance or constrain their participation in any way. The guidelines aren't a punishment, but a set of aspirations—a plan for how each person can get the most out of the meeting's interactions and meaningfully contribute to them. Next we'll see how all of the elements you've planned with such care work together.

Conduct the Meeting

Conduct the Meeting

With your prep work behind you, it's time to transform those efforts into a great meeting with tangible accomplishments. When you log off after leading a productive discussion or resolving a difficult problem, you'll feel exhilaration and relief. But to get to that result, you'll need to be "on" in the meeting: mediating conflict, clarifying misunderstandings, enforcing rules, and maintaining the group's momentum. It's hard work. In this chapter, you'll learn best practices for each phase of the interaction.

Just before the meeting starts

For you, this meeting starts early—by at least 15 minutes. As with a traditional appointment, you'll prepare your physical space to maximize participants' comfort and communicativeness, and then help everyone settle in as they arrive.

Set up the room. Do this whether you're dialing in alone or hosting a handful of attendees in person. Clear your workspace of distractions, and set out the materials you need to run the meeting. If others are joining you, rearrange chairs, check the room's temperature, make space for them to open a laptop, a notepad, or any other items they'll need. If you're all using the same phone to attend the meeting, position that device where everyone can easily hear and be heard.

Video meetings may require some additional tinkering: Does the camera capture everyone in the room? If anyone is backlit or weirdly foregrounded, looming over the rest of the group or over the lens itself, reposition the camera. Pick an angle that minimizes visual distractions for other viewers. For example, if there's a window behind you with a lot of activity on the other side, choose a different background.

Log in early. Whether you're doing a teleconference or a phone call, open up the line a couple minutes ahead of schedule, and hang around while people filter in. If it's a small group, use this opportunity to model the kind of collegiality you want to see during the meeting itself: "Hi, Desmond! How was the game this weekend?" "Maria, looks like a gray day there. What's the weather been like?" You don't have to make brilliant conversation—just the small talk you would if you were all gathering in one room. Not only will these pleasantries set the tone, they also give your attendees the opportunity to casually interact without taking time out of the meeting itself.

With bigger groups, this kind of exchange is impracticable, especially on a conference call or over chat. Instead, consider planning a short warm-up that people can do silently when they join the meeting. Since people often don't read prep materials ahead of time, start the meeting with some quiet reading time, where members spend a few minutes reviewing the material you shared over e-mail. Or use the time to share updates: Before the meeting, ask everyone to send in a sentence about their work, and then share the list digitally right before the meeting starts, so that people can view, add their own updates, and discuss it while stragglers trickle in.

Open the meeting

The first moments of a meeting are powerful, so don't squander them on a litany of logistics. True, you need to communicate a lot of information before you can get to the meat of the meeting. But be thoughtful about how you organize your approach.

Start with purpose. Reaffirm your shared goals. What are you trying to accomplish together? What motivations do you share? Raising these themes at the outset builds a sense of urgency and frames the discussion that follows.

Set the tone. As you outline your goals for the meeting, it's OK to characterize your emotional state openly. Are you optimistic and excited about a project? Frustrated but determined to muddle through? Say so. And while you want to lead with the positive, you also want to be genuine. If you're feeling demotivated, chances are that other people are, too. By acknowledging these emotions, you can shape their effect on the rest of the conversation. So, when you must express ambivalent or negative emotions, adopt a future-oriented, active mind-set: "This situation is so difficult . . . but I believe there are concrete actions we can take to make it better."

Connect people. Ideally, you've fostered some informal chitchat between participants before the meeting officially started. Now, carve out some time so that the participants can identify themselves and explain their meeting roles (presenter, note taker, and so on). These IDs are an easy way to build rapport before you move into weightier matters, and they give everyone a chance to participate in the conversation from the start.

Reiterate key information. Go over the agenda briefly ("We'll start by revisiting last week's sales items, and then Maria will present a draft of her report for our input"), or ask an attendee to describe the materials they've shared ("Can you summarize the recommendation you've made in a sentence or two?"). Then hit the most important process and etiquette points you already shared with your advance materials—in less than a minute, if you can. "Remember, no multitasking, and log your questions on the chat channel so Nicole can facilitate the discussion. Let's try not to interrupt each other, but if we're going long or the conversation gets a little hairy, Mahmoud or I will jump in." Don't recite the whole code of conduct—just what matters most to you.

Get going. Keep the introductory comments outlined above to a minimum, especially if you're the only one talking. Otherwise, folks can easily zone out if it seems like nothing is happening. Unless you must address major issues, such as a new crisis, you should be into your first agenda item within a few minutes.

Facilitate the discussion

In virtual meetings, as in traditional ones, active moderation is everything. Let people ramble on, and you lose the thread of the conversation—and the group's attention. But cutting people off too soon can suppress engagement and damage morale. To find the sweet spot, actively enforce time boundaries *and* solicit group input. (See the sidebar "Sample Language for Facilitators" for more ways to gracefully shepherd the conversation.)

Stay on track. Your agenda allots a certain amount of time to each item. Stick to this plan. If an important, time-sensitive issue is more complex than you anticipated and requires additional discussion, you may have to create more time for this issue by offloading another topic to a later meeting or scheduling a follow-up.

Record action items and tangents. When new, related issues emerge from a discussion of items on your official agenda, track them, as well as any other tangents that come up using an *action-item list*. Document unanswered questions, ideas that weren't pursued, and unresolved disagreements, as well as any participant promises ("I'll report back to the team with those numbers after the call"). Consider making this document public, viewable by other attendees during the call. They can add their own items and see their contributions validated even though you had to redirect the conversation back to the agenda in the meeting itself.

That validation is important, because participants may have a hard time hearing "Not relevant! Moving on!," especially when these messages are stripped of any friendly physical cues. If participants become angry or anxious, however, address those feelings directly before they derail the whole conversation. Capturing their concerns and having them agree that you've documented them appropriately can help move the discussion forward.

SAMPLE LANGUAGE FOR FACILITATORS

Over time, you'll develop a language that keeps your meeting moving. Here are some helpful phrases for specific situations.

- *When the group is silent.* Are they silent because they're listening intently, or because they're confused? Without a visible sea of furrowed brows before you, pause and ask a direct question: "Are you all with me so far?" Or, "I want to take a quick pulse-check. Is everyone following along, or is there anything I can clarify before we move on to the next item on our agenda?"

- *To recover the group's focus after a tangent.* Related but sometimes off-track discussions happen. Recapture everyone's focus by noting the tangent, and getting back to your agenda: "Let's table this point for a moment. I want to return to Diego's comment earlier, about the methodological problems we're facing. Anyone have a response to that?"

- *For someone who keeps reiterating a point.* Acknowledge their feelings, but challenge them to come up with a resolution: "You seem concerned about this decision. What do you think we haven't addressed?"

- *When multiple people are trying to jump into the conversation.* Take the lead and manage who speaks: "Let's finish hearing from Anand, and then we'll hear from Jean."

- *When you catch someone multitasking.* Refocus their attention by calling on them directly: "Sayid, can you chime in here?"

- *When there's excessive background noise.* "Sorry, but I'm having trouble hearing. Mara, can you go on mute until things settle down over there?"

Hear from everyone

Each attendee has something to offer to the conversation—that's why you invited them. But the dynamic between the in-person and remote attendees can worsen conversational imbalances. Quiet people withdraw even further without clear social cues, while blowhards use the lack of feedback as a license to talk over everyone. If you're leading a mixed meeting, the attendees in the room may carry on their own conversation quite happily, leaving no way for remote folks to jump in.

Create a healthy, open dynamic by posing directed questions and gently curbing overparticipation. Build in regular pauses for group input, calling out attendees or locales by name: "Folks in the New Delhi office, are you on board with this idea? Shelly, can you take the room's temperature?" For people whose participation is slightly outside the norm, however, you may need to tailor your tactics.

With quiet people, draw on their existing knowledge with specific questions. Address them by name, so they're sure of who you're talking to, and try prompts such as:

- "Anyara, what has your experience been with this issue?"

- "Will, you've had the longest relationship with this client. Have you encountered this problem with it before?"

- "Suki, I think you went through something similar to this on your last project. What worked for you?"

When someone is dominating the conversation, acknowledge that person's point of view, then pivot to a new topic or question:

- "I want to get back to this at the end of the meeting if we have time. For now, let's focus on X."

- "I hear your point. Does anyone else have a response to it?"

- "We've gotten good input from the yea side of this question. Can we hear from the nays?"

If someone repeatedly talks out of turn, interrupting others and throwing off the group's rhythm, be blunt but polite:

- "Let's talk one at a time. Alice, you were saying?"

- "Please hold questions until the end of the presentation."

- "That's off topic. Right now, we're concerned with X."

Control the tone

As you monitor the flow of ideas, keep an eye on the flow of emotions. Are group interactions civil? Do people feel good about what's happening in the meeting, or do they feel left out, skeptical, and angry? These questions matter because no group can do good work in a bad mood. If people feel snubbed, disrespected, or panicky, they won't contribute meaningfully in the moment, and they may not support your plans afterward. There are several ways to keep the conversation productive.

Listen for tonal shifts. Sometimes, escalating behavior, such as yelling, makes it obvious that the mood of the group is changing. But you'll have to listen for other cues, too, such as when an enthusiastic participant gradually falls silent or a normally laid-back participant suddenly becomes extremely invested in an argument. Did someone abruptly concede a point ("Well, whatever you say")? Look for anomalies— moments where people seem to deviate from what you'd expect from them.

When you notice something unusual, ask the person about it directly: "You seem very concerned about the project timeline. What's driving that?" or "You've been fairly quiet. What's going on in your mind right now?" You might not get a reassuring response. But whatever is causing the behavior change, you risk doing more damage by leaving the issue unaddressed than if you bring it out into the open.

Check out-of-bounds behavior. When someone violates group norms, call them out on it. You don't have to be aggressive ("I can't *believe* you said that!"), but you should head off bad behavior early and say something. Name the problem, then quickly redirect: "Jeff, you're interrupting Gabriel. Gabriel, can you repeat that comment?" You owe it to the rest of the group to uphold the ground rules you set at the beginning of the meeting. This recommendation applies to everything from antisocial behavior, such as bullying, to simple obliviousness, such as exceeding time on a presentation.

Keep calm. Things will go wrong. The unexpected will happen. Don't let these incidents rattle you. Your mood is contagious: If you become agitated or upset, the rest of the group will, too. But if you can shrug off awkwardness and interruption, you'll put everyone at ease.

If you need to ask someone to fix a problem, go through a private channel if possible, such as text or chat, and help the person save face: "Your microphone is acting up, amplifying background noises like your breathing. Is there anything you can do on your end to fix that?" If you must say something while the whole group listens in, choose your words to minimize the embarrassment to your colleague: "I want to remind everyone that while we want to keep the mute button off, if you're experiencing background noise that might distract other folks, go ahead and silence your mic."

Close the meeting

The best way to close a meeting is on time. Still, you want to use these last few moments for more than "Thanks! Till next week." To double-down on your team's alignment and supply a powerful sense of momentum for next steps, use the following tactics.

Agree on next steps. Ask the group, "What will we do by our next meeting to ensure progress?" Lay out the specific commitments each member of the group is making. Create a list of tasks or responsibilities, attached to a name and a timeline. "Heba is going to put together a budget for this project list by March 1, and then Damian and Thien will start working up a prototype for our next meeting on March 15." It's crucial to have this conversation during the meeting itself. Working at a distance makes it significantly more difficult to follow up with people who don't particularly want to make themselves available (more on that in a moment), but a verbal commitment activates a sense of obligation in most people.

As you're making plans, periodically check to make sure that everyone is onboard with the group's conclusions. Ask direct questions, maybe even person by person:

- "Is there anything else anyone needs to say or ask before we finalize this plan?"

- "Are you comfortable with where we ended up?"

- "What would it take for you to be OK with this?"

Articulate the value of what you've accomplished. Just as you opened your meeting by reviewing your shared purpose, close by reminding them why this work is important: "We've planned a recruitment strategy for key company positions. Filling those roles will make a big difference in the company's stability and in our ability to carry out initiatives dear to all of our hearts. Great job, all!"

Technology makes it easy for people to do a slow fade-out, so at this point your colleagues might already be shifting their attention, waiting for the cue to say good-bye and hang up. Don't let that happen— you'll lose out on a crucial opportunity to fortify their motivation. Instead, flip the script and invite *them* to give this benediction. For example, ask each person to say one thing the meeting has done for them: "I'm going to be thinking about Jason's presentation as I prepare for my next client meeting" or "I'm glad I got to air my questions about the spring exhibition and help shape our plan for the next phase of this project."

Follow up

E-mail the participants shortly after your meeting (ideally, no later than the next day), just as you would for an in-person meeting. But with virtual meetings, you can't stop by participants' desks or catch them on the way to the break room, so this e-mail makes your expectations visible and helps hold people accountable. Document and share the specific decisions and outcomes you arrived at, note who's responsible for follow-up tasks, and clarify when the tasks must be completed. Figure 2 shows a sample follow-up note. Include a copy of the scribe's notes or a recording of the meeting, if you have it. Then, check in (repeatedly, as necessary) through different channels such as text or IM or a casual phone call until everyone has met their commitments.

In general, keep your tone positive so it doesn't sound as if you're micromanaging them, and close with an offer to help: "Do you need anything from me to complete this?" If they're unresponsive or don't seem to be making progress, your best bet is to own your concern openly. Distance makes this sort of directness necessary, but you can lessen the negativity by offering support. If you know that this person has a lot of other stuff going on right now, acknowledge

FIGURE 2

Sample follow-up note

Special Revenue-Generating Project: Reprioritizing Workloads

Follow-up notes from January 25, 2016, meeting
Attendees: Emily, Lisa, Jack, Chris, Angela, and Steve
Meeting objective: To develop plan for shifting other responsibilities of key team members

In this meeting, we discussed approaches to lighten Chris's and Angela's workloads. We have determined that the best approach will be to ask Steve's direct report Sarah to take on several of their responsibilities.

Next steps:

- Steve will discuss this change with Sarah by next Wednesday and confirm that she has the bandwidth to do the extra work.
- Once Sarah is made aware of her added responsibilities, Lisa will send an e-mail to the full group and add her to the team Slack channel, and Emily will send an e-mail to the Special Revenue-Generating Project team, updating it about the changes.
- Jack will inform the group about no longer sending out sales reports.
- Chris and Angela will set up individual meetings with Sarah to train her on their responsibilities.

What	How	Who	When
Opening	• Importance of Special Revenue-Generating Project		
Team members' current responsibilities	• Role of key team members on the project • Compiling and sending out weekly sales reports; making sure marketing has information it needs about upcoming products; main point of communication with two key sales accounts	Chris	

What	How	Who	When
Possible responsibilities to shift away from team members	• Investigating sales leads for a critical product line; sales specialist for that product line; main point of communication with two other key sales accounts	Angela	
	• Can't move key accounts responsibilities—these are long-standing relationships		
Potential solutions and approaches	• Sarah has been asking for more responsibility—could give her Angela's role on the other product line	Angela	
	• Could stop sending out sales reports since they are all available on corporate intranet now—need to advise group	Jack	Next Friday
Decision on an approach	• Steve is concerned about adding all of this to Sarah's plate—need to talk to her	Steve	Next Wednesday
	• Team votes using Doodle poll and decides that this is still the right approach; Jack approves		
Next steps	• Send e-mails notifying sales team and Special Revenue-Generating Project team of changes	Lisa, Emily	After Sarah is notified
	• Add Sarah to the team Slack channel	Lisa	After Sarah is notified
	• Set up training for Sarah	Chris, Angela	Within two weeks

this generously: "I know you're under a lot of pressure with the board meeting coming up, and I appreciate your willingness to put this on your plate, too. But I'm worried that I haven't received an answer from you to my e-mails and texts last week. What's going on? And how can I help you move forward?"

Leading a successful meeting requires energy and presence of mind. Sometimes this role is great fun, especially when you're working with an engaged, enthusiastic crowd that's playing off one another. Sometimes it's stressful, such as when your colleagues disagree vehemently or when no one seems to care. These are difficult moments, but well within the bounds of normal. Sometimes, however, you'll face more unusual situations. In the next chapter, you'll learn how to deal with two challenges that are slightly beyond the scope of regular virtual meeting logistics.

Navigate Special Situations

Navigate Special Situations

With a little bit of practice, you'll soon feel like a natural leader in the virtual environment. You may still experience occasional awkwardness and confusion, but as your repertoire of responses grows, you'll feel more capable of putting others at ease and running productive meetings. But occasionally, some situations require extra care. In this chapter, you'll learn how to handle yourself in two of these different and challenging scenarios.

Give a video presentation

You've been asked to give a big presentation to a scattered audience. That's great—but also potentially nerve-wracking. Before you get caught up in the technological aspects of your talk, spend some time on the basics. Video presentations, like live ones, need clear goals and a strong core message. Focus what you'll say by asking yourself questions such as these:

- What's the purpose of my presentation—for example, am I teeing up a specific decision or problem?

- Who's my audience? Who else will be attending the meeting or reviewing my materials after the fact?

- What platform will the meeting use? What communication technologies or presentation apps are available to me?

- How much time do I have?

- Who's the technology point-person for this meeting, and how can I get in touch with them? Will they need anything from me before the meeting—a copy of my presentation, any specs or contact information?

As you're preparing the presentation itself, keep the graphics lively, and avoid densely written slides.

Once you've drafted your remarks and the slides that will accompany them, practice—a lot. Time yourself, leaving room for audience participation, and do at least one dry run with the technology you'll be using during the meeting itself. For example, if you're presenting through a teleconference app, make sure you know how to navigate its presentation functions. If you'll be visible to the audience on a video screen, experiment with what you'll wear and where you position the lens, so that you don't discover at the last minute that your plaid shirt breaks down on screen into wildly distracting pixelated waves or that your eyes look glazed when you read through your speaker's notes.

When you begin your presentation, tell the other participants how you'd like to handle questions and comments. It's awkward to stop your words midstream, especially over the phone, where people can't signal their intentions with questioning facial expressions or eager body language. So, ask participants to wait until a Q&A at the end, to keep track of questions in a shared Google Docs file, or to feed you questions via chat or text while you're talking. If you're open to interruptions, explain how the attendees can get your attention.

Throughout the presentation, keep your audience engaged by drawing them in. If you're using a technology that allows for other attendees to be heard, periodically solicit their input by posing questions ("What's the first thing that jumps out at you when you look at this graph?") or hat-tipping experts ("This data is from Chandra's group—thank you! Can you quickly explain how your team sourced it?"). If the attendees can't jump into the discussion, you can still pose questions to get them thinking. Pause for a second before you reveal the answer. If they came up with the "right" one, they'll feel satisfied and smart. If not, they'll feel surprised—and want to understand why.

Finally, don't exceed your allotted time. If you haven't gotten through everything you planned, wrap things up as gracefully as you can ("This is a good stopping point. Please follow up with me if you have more questions"), or ask the meeting leader if you can have more time.

Bridge linguistic or cultural barriers

If you're crossing global boundaries to bring together people with different languages and cultural norms, you may feel additional pressure to make their interactions work. The danger of miscommunication is even higher than normal (and in a virtual setting, it's always fairly high).

Before you implement any particular solutions, think about the barrier you're dealing with. Some people will have a difficult time understanding the speech of someone who doesn't share their native language if they can't watch the person's face. Participants who don't understand what's being said might interrupt constantly to clarify. Or they might simply fall silent, too embarrassed to reveal that they can't follow what's going on.

But understanding language goes beyond the simple recognition of sounds. Your participants might have fundamentally different ways of communicating about their priorities. In cultures that prize direct language, it's customary to say, "That deadline is too late. We need to move it up." In cultures that communicate with more indirect language, someone might choose to say instead, "Thank you for creating this schedule. I am very eager to see it completed."

In addition, it's important to recognize that all of our norms around the use of technology are culturally conditioned. For some of your colleagues, it might be acceptable to enter and exit a room while a teleconference is going on. Others might view this as the height of rudeness. Fortunately, you don't have to resolve these big cultural gaps. You just have to clarify which behaviors are OK in your own meeting, and help people live up to them.

Ask participants how they prefer to communicate. Make this part of your technology-prep protocol (see the earlier chapter "Manage the Technology"). Some people might find it easier to understand another speaker when they can use a landline or when they can watch other people's faces as they talk on video. Accommodate these preferences, if you can. Your colleagues will be more confident and focused during the meeting if they find that the tech setup doesn't put them at a disadvantage.

Identify important cultural norms. You can't—and shouldn't—shed all your cultural habits, but with a little research, you can learn a new behavior that will build goodwill and avoid a basic cultural offense. Pay attention to things such as how to welcome and address attendees (even virtual ones) and the social etiquette of saying hello, good-bye, and thanks. For these matters, it's better to talk to someone with experience than to run a Google search. Tap your network to find someone familiar with both the culture of your collaborators and your particular work environment.

Find common ground—ahead of time. Research shows that similar life experiences or a shared background, however minor, is your best bet for building trust quickly. Use Google, LinkedIn, Twitter, Facebook, and any other platform that comes to mind to learn about the people you've invited. Where did they go to school? Do they have kids? Share related hobbies? And so on. When you open the meeting, include these details in a thoughtful round of introductions. Trust grows over discussing even such small things.

Periodically summarize the conversation. If people are consistently struggling to understand one another's speech, it becomes tiresome to ask for clarification. Compensate by kicking your moderating skills into a higher gear. After each comment or question, briefly paraphrase what was said. For example:

- "That's an important question. I agree that the production facility's capacity is a major risk in this plan. Here's what we know . . ."

- "So in your experience, this research technique requires a lot of resources, maybe more than we have. What does everyone else think about that?"

Follow up with extra care. Documenting your agreements and following up will give the participants time to fully absorb what all of you covered and will help avoid miscommunication. If a colleague is freezing you out and shirking their follow-up assignments, find someone culturally proficient to help you interpret your communications. You might need to identify a better strategy if you discover that there may be something happening other than the person's being overloaded or disinterested in the assigned task.

. . .

Running a virtual meeting requires focus, tact, and discipline. But for all the ways a meeting can go off the rails, there are so many more ways it can go right. When you call on your quietest collaborator and they resolve a raging disagreement with one well-placed comment. When everyone bursts into laughter at the same weird, high-pitched noise, and your glitchy tool becomes the group's new in joke. When you end a call by telling your coworkers to enjoy the weekend—and they wish you a happy Friday morning. When two colleagues in different cities, both speaking a second language, successfully diagnose a problem for the rest of the team. Your meeting doesn't need to be perfect to be a success. With a bit of planning and an alert spirit, you can carry anything off.

Learn More

Quick hits

Frisch, Bob and Cary Greene. "Before a Meeting, Tell Your Team That Silence Denotes Agreement." HBR.org, February 3, 2016. https://hbr.org/2016/02/before-a-meeting-tell-your-team-that-silence-means-agreement.

Virtual meetings give attendees an excuse to withhold their active participation and reframe themselves as observers. But when people conceal their opinions in the moment, they're unlikely to buy into the outcome of a discussion afterward. Frisch and Greene show you how to thwart this dynamic. They suggest that you tell your colleagues that silence has a clear meaning, and they offer tips to increase engagement, from administering polls to breaking into smaller groups.

McKee, Annie. "Empathy Is Key to a Great Meeting." HBR.org, March 23, 2015. https://hbr.org/2015/03/empathy-is-key-to-a-great-meeting.

To compensate for the loss of context in a virtual setting, your emotional intelligence must work overtime to read the group and manage its emotional dynamics. McKee explains

how empathy helps you understand how participants are re-
lating to each other and why staying attuned to other people's
feelings may require you to regulate your own more effectively.

Samuel, Alexandra. "Digital Tools to Make Your Next Meeting More Productive." HBR.org, July 3, 2015. https://hbr
.org/2015/07/digital-tools-to-make-your-next-meeting-more
-productive.

Samuel walks you through a handful of tools that can help
you map ideas, engage participants, tweak processes, and turn
goals into concrete tasks. Illustrated with screenshots from a
variety of tools, this article will help you select something new
to try at your next meeting.

Books

Duarte, Nancy. *HBR Guide to Persuasive Presentations*. Boston: Harvard Business Review Press, 2012.

To improve your virtual presentations, first master the basics. Presentation expert Duarte shows you how to win over
tough crowds, organize a coherent narrative, create powerful
messages and visuals, and more. This book takes a big-picture,
strategic look at a common activity and breaks it down into
practical action steps.

Harvard Business School Publishing. *HBR Guide to Making Every Meeting Matter*. Boston: Harvard Business Review
Press, 2011.

Running any type of meeting effectively requires preparation. But it doesn't have to be painful. Drawing from multiple HBR contributors, this guide offers a range of tips to make your meetings easier to prepare for, more enjoyable to run—and more productive.

Molinsky, Andrew. *Global Dexterity: How to Adapt Your Behavior Across Cultures Without Losing Yourself in the Process.* Boston: Harvard Business Review Press, 2013.

To facilitate virtual meetings across cultures, you have to be open to new ways of doing things. But effective leadership depends on authenticity—on your ability to speak to, and act on, your real values and beliefs. Molinsky expounds a new approach to this problem. His study of global dexterity shows readers how to try out new behaviors when crossing cultures without triggering emotional and psychological barriers.

Articles

Conger, Jay. "The Necessary Art of Persuasion." *Harvard Business Review*, May 1998 (product #98304).

In a virtual environment, it's hard to tell whether your ideas are landing accurately. Since you can't rely on implicit feedback to tighten a pitch or to refine an argument, you need to come prepared to persuade. Conger's seminal article explains how persuasion really works—it's a process of negotiation and mutual learning, not a hard sell—and describes how four best practices, ranging from establishing credibility to framing common ground, will help you reach even the most pixelated colleague.

Halvorson, Heidi Grant. "A Second Chance to Make a First Impression." *Harvard Business Review*, January 2015 (product #R1501J).

When you lead a virtual discussion, you're constantly managing other people's perceptions—of you, of each other, of the business at hand. Halvorson dissects how human perception works and what biases are at work when we make judgments about one another. Her advice on how to leverage these biases to your own advantage will help you plan your meeting strategy and mend fences between unhappy collaborators.

Leonard-Barton, Dorothy and William A. Kraus. "Implementing New Technology." *Harvard Business Review*, November 1985 (product #85612).

This classic article examines the problems you face when you ask meeting attendees to adopt new communications technology. Your colleagues look to you to explain how this innovation works—and why they should bother to use it. Although the technology has changed since this article was published, the problems it describes still apply today. By viewing yourself as an internal marketer, as the authors suggest, you'll have better luck getting buy-in for new platforms and overcoming resistance to change.

Sources

Primary sources for this book

Ferrazzi, Keith. "How to Run a Great Virtual Meeting." HBR
.org, March 27, 2015. https://hbr.org/2015/03/how-to-run
-a-great-virtual-meeting.

Harvard Business School Publishing. *Running Meetings*
(20-Minute Manager Series). Boston: Harvard Business
Review Press, 2014.

Other sources consulted

Axtell, Paul. "The Right Way to End a Meeting." HBR.org,
March 11, 2015. https://hbr.org/2015/03/the-right-way-to
-end-a-meeting.

Clark, Dorie. "The Right (and Wrong) Way to Network." HBR
.org, March 10, 2015. https://hbr.org/2015/03/the-right
-and-wrong-way-to-network.

Duarte, Nancy. "Five Presentation Mistakes Everyone Makes."
HBR.org, December 12, 2012. https://hbr.org/2012/12/
avoid-these-five-mistakes-in-y.

Sources

Ferrazzi, Keith. "Five Ways to Run Better Virtual Meetings."
HBR.org, May 3, 2012. https://hbr.org/2012/05/the-right
-way-to-run-a-virtual.

Gallo, Amy. "The Seven Imperatives to Keeping Meetings on
Track." HBR.org, December 20, 2013. https://hbr.org/
2013/12/the-seven-imperatives-to-keeping-meetings
-on-track.

Gino, Francesca. "Introverts, Extroverts, and the Complexities
of Team Dynamics." HBR.org, March 16, 2015. https://hbr
.org/2015/03/introverts-extroverts-and-the-complexities
-of-team-dynamics.

Harvard Business Review Staff. "A Checklist for Planning
Your Next Big Meeting." HBR.org, March 26, 2015.
https://hbr.org/2015/03/a-checklist-for-planning-your
-next-big-meeting.

———. "How to Know if There Are Too Many People in
Your Meeting." HBR.org, March 18, 2015. https://hbr.
org/2015/03/how-to-know-if-there-are-too-many-people-
in-your-meeting.

Harvard Business School Publishing. *Leading Virtual Teams*
(Pocket Mentor Series). Boston: Harvard Business Press,
2010.

McKee, Annie. "Empathy Is Key to a Great Meeting." HBR
.org, March 23, 2015. https://hbr.org/2015/03/empathy
-is-key-to-a-great-meeting.

Molinsky, Andy and Melissa Hahn. "Learning the Language
of Indirectness." HBR.org, May 6, 2015. https://hbr.org/
2015/05/learning-the-language-of-indirectness.

Morgan, Nick. "How to Conduct a Virtual Meeting." HBR.
    org, March 1, 2011. https://hbr.org/2011/03/how-to-con-
    duct-a-virtual-meeti.
Rousmaniere, Dana. "What Everyone Needs to Know About
    Running Productive Meetings." HBR.org, March 13, 2015.
    https://hbr.org/2015/03/what-everyone-needs-to-know
    -about-running-productive-meetings.
Saunders, Elizabeth Grace. "Do You Really Need to Hold That
    Meeting?" HBR.org, March 20, 2015. https://hbr.org/
    2015/03/do-you-really-need-to-hold-that-meeting.

Index

Notes

Notes

Notes

Notes

Notes

Notes

Smarter than the average guide.

Harvard Business Review Guides

If you enjoyed this book and want more comprehensive guidance on essential professional skills, turn to the **HBR Guides series**. Packed with concise, practical tips from leading experts—and examples that make them easy to apply—these books help you master big work challenges with advice from the most trusted brand in business.